Comets

by Joanne Mattern

FOCUS READERS

BEACON

www.focusreaders.com

Focus Readers is distributed by North Star Editions:
sales@northstareditions.com | 888-417-0195

Produced for Focus Readers by Red Line Editorial.

Photographs ©: ESA/Rosetta/NAVCAM/JPL/NASA, cover, 1, 10, 29; J. da Silva/NSF/AURA/NOIRLab, 4; P. Bernardinelli & G. Bernstein (UPenn)/DESI Legacy Imaging Surveys/Dark Energy Survey/DOE/FNAL/DECam/CTIO/NSF/AURA/NOIRLab, 7; NASA Goddard/GSFC/NASA, 8, 13; Fine Art Images/Heritage AiWire/Newscom, 14; ESA/Giotto Project/JPL/NASA, 17; Larry Mayer/Billings Gazette/AP Images, 19; iStockphoto, 20–21; JPL-Caltech/UMD/NASA, 22; KSC/NASA, 25; ESA/JPL/NASA, 27

Library of Congress Cataloging-in-Publication Data
Library of Congress Cataloging-in-Publication Data is available on the Library of Congress website.

ISBN
978-1-63739-245-4 (hardcover)
978-1-63739-297-3 (paperback)
978-1-63739-399-4 (ebook pdf)
978-1-63739-349-9 (hosted ebook)

Printed in the United States of America
Mankato, MN
082022

About the Author

Joanne Mattern has written many nonfiction books for children. Her favorite topics are science, sports, history, and animals. Joanne enjoys showing young readers the fascinating people, places, and things in our world. She lives in New York State with her family. She enjoys doing jigsaw puzzles and taking long walks.

Table of Contents

Chapter 1

Big and Bright

In 2021, two **astronomers** looked at pictures from deep space. They were surprised to see a big, bright comet. It was millions of miles away. It was one of the largest comets ever seen.

This illustration shows Comet Bernardinelli-Bernstein. It is named after the astronomers who found it.

The comet was headed toward the Sun. That made astronomers excited. They would be able to study it for years. Astronomers had never studied such a distant comet as it approached the Sun. They hoped to learn how comets change over time.

Did You Know?

Comets are usually named after the people who discover them.

This photo shows the comet (circled) as it appeared to astronomers in 2021.

Astronomers believed the comet would get closest to the Sun in 2031. After that, it would go back into deep space.

Chapter 2

Space Snowballs

Comets are objects in space. They **orbit** the Sun. Comets are made of ice, rock, dust, and frozen gases. These materials are very old. They are left over from when the **solar system** formed.

Solar systems form when huge clouds of gas and dust are pulled together by gravity.

The nucleus of Comet 67P is approximately 2.5 miles (4.0 km) across.

A comet has several parts. The solid **core** is called the nucleus. On most comets, the nucleus is not very big. It is usually less than 6 miles (10 km) across.

During part of its orbit, the comet moves toward the Sun. It gets closer and closer. The Sun's heat causes the nucleus to warm up. Some of the ice turns into a cloud of gas. This cloud is called the coma. The coma is large. It can be 1,000 times bigger than the nucleus.

A comet also has two tails. One tail is made of dust. The other is made of gas. Both tails are caused by the Sun. The Sun blows dust and gas off the comet as it approaches. For this reason, the tails always point away from the Sun.

Scientists have found thousands of comets. However, they believe

Did You Know?

A comet's tails can be millions of miles long.

Scientists observed the tails of Comet ISON as it flew near the Sun in 2013.

there are billions of comets in our solar system. Most of them are very far away. They lie beyond the farthest planet, Neptune.

Chapter 3

Looking at Comets

People have been looking at comets since ancient times. Long ago, people thought comets orbited Earth. But in 1577, an astronomer found that comets travel far into space.

Tycho Brahe studied comets in the 1570s.

In the 1680s, Isaac Newton discovered that comets orbit the Sun. Newton's friend, Edmond Halley, studied these ideas. Halley learned that comets appear in the sky again and again.

Halley's Comet is one of the most famous comets. It is named after Edmond Halley. This comet orbits the Sun every 75 years.

In March 1811, an astronomer spotted a new comet. Every day, it got brighter and brighter. By

Halley's Comet was last seen in 1986. It will return in 2061.

October, it was one of the brightest objects in the night sky. Authors wrote about it. Artists put it in their paintings. It became known as the Great Comet of 1811. Its orbit takes much longer than Halley's Comet.

It will not come back for thousands of years.

A comet called Hale-Bopp arrived in 1996. It shone in the sky for months. In 1997, it made its closest approach. It flew within 124 million miles (200 million km) of Earth. This comet had a large, bright nucleus. People could see this

Did You Know?

In 1994, a comet hit Jupiter. Scientists on Earth watched the huge explosion.

In March 1997, Hale-Bopp was easy to see in the night sky.

comet without **telescopes**. It was brighter than the stars. Scientists were happy to study a comet so close to Earth.

THAT'S AMAZING!

Scary Comets

Long ago, people feared comets. They thought comets were signs of doom. Some people thought the Great Comet of 1811 was linked to a huge earthquake. Others connected the comet to wars.

A few people were still afraid of comets in 1910. That year, Halley's Comet returned. Some people sealed their windows shut. They worried that the comet would cover Earth with poison gas. Today, we know there is nothing to fear from comets.

The Great Comet of 1811 was visible for more than eight months.

Chapter 4

Exploring Comets

In 2005, scientists crashed a **probe** into a comet. They studied what was inside of the comet. Cameras filmed the explosion. For the first time, scientists saw the inside of a comet's nucleus.

A probe explodes after smashing into Comet Tempel 1 in 2005.

In the late 2000s, scientists studied **samples** from another comet. A spacecraft gathered bits of gas and dust. Then the spacecraft brought these samples back to Earth. Scientists were amazed. They found a **protein** called glycine. This protein is an

Did You Know?

Billions of years ago, many comets crashed into Earth. They left behind chemicals. Some scientists think these chemicals helped life begin.

Scientists spent years building the Stardust spacecraft. It collected samples from a comet.

important part of living things. It was the first time scientists had found glycine in a comet.

In 2014, scientists sent a spacecraft into orbit around a comet. Then, a probe landed on the comet's surface. Scientists noticed that the comet made strange **vibrations**. Studies showed the comet might have a hard outside and a soft inside.

Scientists planned to launch a new **mission** in the late 2020s. Three spacecraft will study a comet as it travels toward the Sun. The spacecraft will take many pictures.

An illustration shows the Philae lander on the surface of Comet 67P.

These images will help scientists figure out what the comet is made of. Scientists are always learning more about comets.

FOCUS ON
Comets

Write your answers on a separate piece of paper.

1. What is the main idea of Chapter 3?

2. Would you want a job studying comets? Why or why not?

3. When did scientists crash a probe into a comet?

- **A.** 1811
- **B.** 1997
- **C.** 2005

4. Scientists think there are billions of comets. If this is true, why don't we see comets very often?

- **A.** Most of them are invisible.
- **B.** Most of them are very far away.
- **C.** Most of them move too fast.

5. What does **approaches** mean in this book?

The Sun blows dust and gas off the comet as it ***approaches****. For this reason, the tails always point away from the Sun.*

A. moves away from something
B. moves toward something
C. moves below something

6. What does **ancient** mean in this book?

People have been looking at comets since ***ancient*** *times. Long ago, people thought comets orbited Earth.*

A. scary and confusing
B. new or modern
C. far in the past

Answer key on page 32.

Glossary

astronomers
Scientists who study space.

core
The central part of something.

mission
An important task or operation.

orbit
To repeatedly follow a curved path around another object because of gravity.

probe
A device used to explore.

protein
A substance that tells a living cell what to do.

samples
Small pieces taken from a larger object for study.

solar system
The Sun and all the objects that move around it, including planets, moons, asteroids, and comets.

telescopes
Tools that allow people to see faraway objects.

vibrations
Very quick back-and-forth movements.

To Learn More

BOOKS

Harman, Alice. *Asteroids and Comets.* New York: Rosen Publishing Group, 2021.

Kurtz, Kevin. *Comets and Asteroids in Action (An Augmented Reality Experience).* Minneapolis: Lerner Publications, 2020.

Rathburn, Betsy. *Comets.* Minneapolis: Bellwether Media, 2019.

NOTE TO EDUCATORS

Visit **www.focusreaders.com** to find lesson plans, activities, links, and other resources related to this title.

Index

Answer Key: 1. Answers will vary; **2.** Answers will vary; **3.** C; **4.** B; **5.** B; **6.** C